Material Choices

CAN YOU MAKE A BOOK OUT OF METAL?

by Susan B. Katz

PEBBLE
a capstone imprint

Published by Pebble, an imprint of Capstone
1710 Roe Crest Drive, North Mankato, Minnesota 56003
capstonepub.com

Copyright © 2023 by Capstone. All rights reserved. No part of this publication may be reproduced in whole or in part, or stored in a retrieval system, or transmitted in any form or by any means, electronic, mechanical, photocopying, recording, or otherwise, without written permission of the publisher.

Library of Congress Cataloging-in-Publication Data is available on the Library of Congress website
ISBN: 9781666350876 (hardcover)
ISBN: 9781666350937 (paperback)
ISBN: 9781666350999 (ebook PDF)

Summary: Books are made with paper. But what if we tried to make a book with metal? Copper, tin, gold, silver—would any of those materials work to make a book? Find out more about metal, its properties, and how it is produced.

Editorial Credits
Editor: Christianne Jones; Designer: Elyse White; Media Researcher: Morgan Walters; Production Specialist: Polly Fisher

Image Credits
Getty Images: Images By Tang Ming Tung, 5, Jill King / EyeEm, 11, John W Banagan, top right 7, RHJ, bottom right 7; Shutterstock: domnitsky, left Cover, eremeevdv, 15, Ja Crispy, 13, Kokhanchikov, 19, Lubos Chlubny, 16, Marko Kukic, 9, Meddy Gunawan, top middle 7, pixelheadphoto digitalskillet, 17, RHJPhtotoandilustration, top left 7, bottom left 7, smspsy, 10, somdul, 12, Vladislav Lyutov, 21, xpixel, right Cover

All internet sites appearing in back matter were available and accurate when this book was sent to press.

Printed and bound in the USA. 4882

Table of Contents

A Metal Book? 4
Lots of Metals 6
Finding Metal 8
Properties of Metal 10
Made from Metal 12
Using Metal 14
Metal in Books 18
    Activity 20
    Glossary 22
    Read More 23
    Internet Sites 23
    Index 24
    About the Author 24

Words in **bold** are in the glossary.

A Metal Book?

Books are made for reading. A book has paper pages. Words are printed on pages using ink. It should be light enough to carry and hold. But it needs to be strong enough that it doesn't tear.

So would metal make a good material for making a book? Let's find out more!

Lots of Metals

There are many different metals. Some metals are heavy and hard, like iron and copper. Other metals are light and soft, like aluminum and tin.

Some metals are common. They are easy to find. These metals include iron, steel, and copper. They are cheap metals.

Other metals are **rare**, like gold and silver. They are hard to find, so they cost a lot of money.

Common Metals

aluminum   tin   iron

Rare Metals

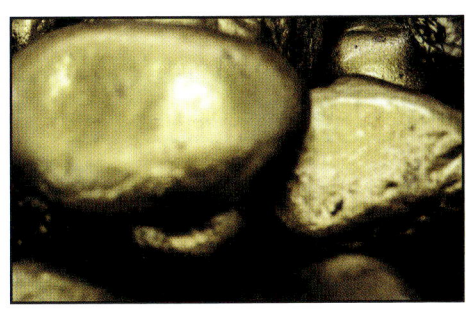 

gold   silver

Finding Metal

Most metals are found in rocks. The rocks come from Earth's crust. Metal has to be taken out of rocks or **minerals**.

You can mix metals to make **alloys**, like steel. Steel is a combination of iron and several other **elements**. Steel is used more than any other metal to make buildings all around the world.

Properties of Metal

Metals are strong and heavy. Most have a high melting point. They can be changed. You can melt or bend them. You can also stretch metal to be very thin. It can be made into different shapes.

Made from Metal

Pots and pans are made out of metal. They trap heat. They also have a very high melting point, so they won't melt on the stove.

The metal frame of a car

Buildings, airplanes, and cars have metal frames. Many frames are made of steel and aluminum because these metals are very strong. They will help protect people in an accident.

Using Metal

Metal can move **electricity** from one object to another. People use metal to build computers. Some of the metals used to make computers are aluminum, cobalt, copper, and iron. Silver is also used to make cell phones, TVs, and keyboards.

The inside of a TV

Many other items are made of metal too.
Jewelry is often made of gold and silver.
It is rare, shiny, and lasts a long time.

Most pipes are made of metal. The metal is shaped into a hollow tube so water can pass through it without leaking. Even braces on teeth are made out of metal!

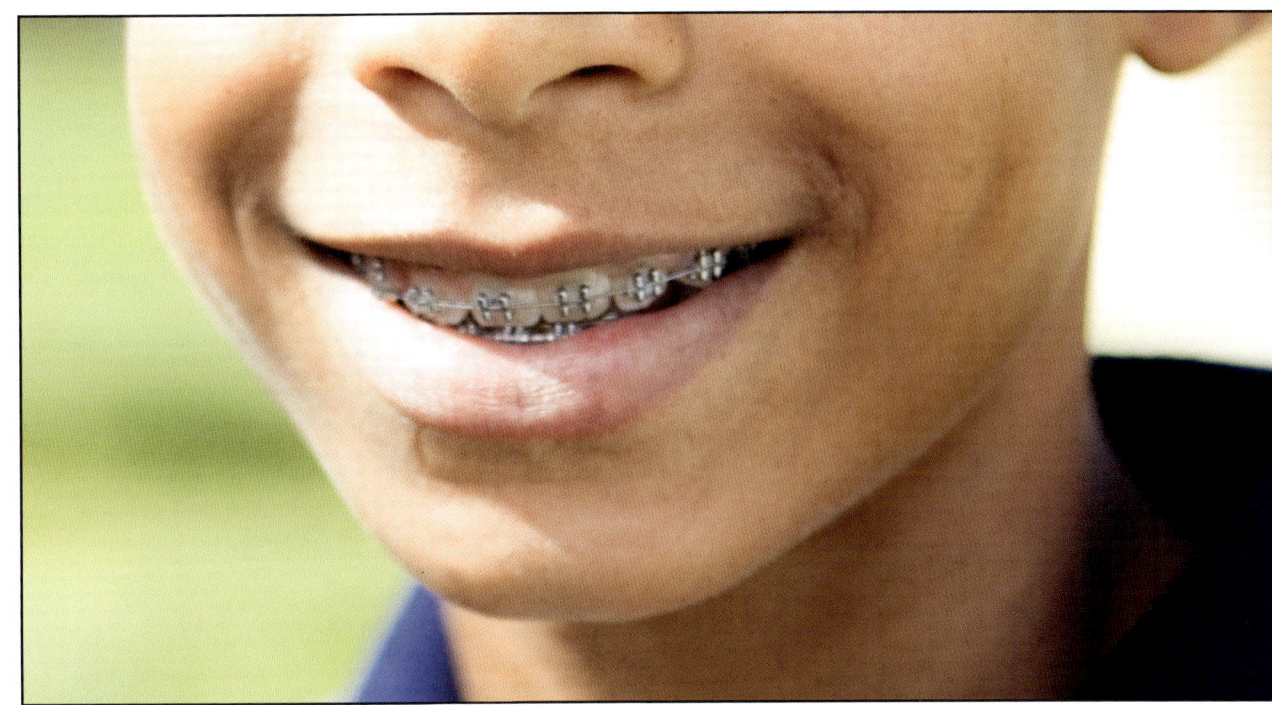

Metal in Books

You could stretch metal to make it thin. Metal pages would be strong. They wouldn't tear easily. But a book needs to be light. A metal book could be very heavy to carry around.

People need to be able to see the words on a page. It is not easy to see words stamped on metal. Reading a metal book would be very hard.

So could you make a book out of metal? Would you?

ACTIVITY

What Would YOU Use?

Books need to be printed, cut, put together, and shipped. What material would you use to make a book? Can you make one out of materials that you have at home? See if you have the items below. Maybe you can find other materials to use as well.

What You Need:

- 4 sheets of white paper
- 1 colorful sheet of paper
- a stapler

What You Do:

1. Fold all the paper in half.

2. Put the white paper inside the colorful paper.

3. Staple the folded side together on the edge.

4. Now write a story and draw pictures to match your story.

5. Be sure to add a title to the front of your book and your name as author.

Glossary

alloy (AL-oi)—combination of two or more metals

electricity (ih-lek-TRIS-i-tee)—a natural force that can be used to make light and heat or to make machines work

element (EL-uh-muhnt)—a substance that can't be split into simpler substances

mineral (MIN-er-uhl)—a solid found in nature that has a crystal structure

rare (RAIR)—not common

Read More

Amstutz, Lisa J. *Is It Shiny or Dull?* North Mankato, MN: Capstone, 2022.

Howes, Katey. *Be a Maker.* Minneapolis: Carolrhoda Books, 2019.

Rivera, Andrea. *Metal.* Minneapolis: Zoom, 2018.

Internet Sites

Britannica Kids: Metal
kids.britannica.com/kids/article/metal/390809

Ducksters: Metals
ducksters.com/science/metals.php

Kiddle: Metal Facts for Kids
kids.kiddle.co/Metal

Index

airplanes, 13
aluminum, 6, 7, 13, 14
buildings, 8, 13
cars, 13
computers, 14
copper, 6, 14
electricity, 14
gold, 6, 7, 16
iron, 6, 7, 8, 14
jewelry, 16
minerals, 8
pans, 12
pipes, 17
pots, 12
rocks, 8
silver, 6, 7, 14, 16
steel, 6, 8, 13
tin, 6, 7

About the Author

Susan B. Katz is an award-winning Spanish bilingual author, National Board Certified Teacher, educational consultant, and social media strategist. When she's not writing, Susan enjoys salsa dancing and spending time at the beach.